Enigma

UNSCRAMBLED

Philip Bauer

ENIGMA UNSCRAMBLED

iUniverse books may be ordered through booksellers or by contacting:

iUniverse
1663 Liberty Drive
Bloomington, IN 47403
www.iuniverse.com
844-349-9409

Because of the dynamic nature of the Internet, any web addresses or links contained in this book may have changed since publication and may no longer be valid. The views expressed in this work are solely those of the author and do not necessarily reflect the views of the publisher, and the publisher hereby disclaims any responsibility for them.

Any people depicted in stock imagery provided by Getty Images are models, and such images are being used for illustrative purposes only.
Certain stock imagery © Getty Images.

ISBN: 978-1-6632-3316-5 (sc)
ISBN: 978-1-6632-3317-2 (e)

Library of Congress Control Number: 2022900306

Print information available on the last page.

iUniverse rev. date: 01/18/2022

CONTENTS

FOREWORD

During World War II the armed forces of Germany used the Enigma cipher machine to encode and decode secret messages. By late 1940 the best British codebreakers had been able to fashion electronic machines which helped them decode the captured radio messages. Today, with minute detailed knowledge of the Enigma and how it worked and the techniques used by the Germans to encode the messages, modern computers can decode those messages in what would seem to be an instant to the 1940 codebreakers. Since then, encoding and decoding techniques utilizing high speed computers have advanced to levels unimaginable in 1940, but even today the general public remains fascinated with all aspects of the Enigma as used in WWII.

My interest in the Enigma developed over a number of years. I finally decided that the best way to understand it would be to simulate it with a computer program and if possible to also write a program to decode a "captured" Enigma message based on assumptions which matched the knowledge the British had in 1940.

I retired from software development some time ago, so I must admit the effort to design, code, and test the simulator — a task requiring some attention to detail —seemed to be an excellent way to see what skills remained.

The following pages describe the Enigma, how it worked, and how it was used. The simulator is described and how to use it. I also discuss my efforts to write a raw message decode program. And for those who are interested in the program itself, my source code for the simulator is listed in the Appendix.

And finally, I wish to thank my son-in-law Matthew Crummey for the professional graphics in Figure 3, *The Example Showing Enigma Functional Wiring.*

Philip Bauer

November 2021

BACKGROUND FOR
SIMULATOR DEVELOPMENT

In the mid-1980's I read *The Ultra Secret* by F.W. Winterbotham which discussed the Enigma code machine used by Germany in WWII. It told how the British had developed ways of analyzing and decoding the German secret messages, and how they had kept this capability a closely guarded secret basically until the mid 1970's. The author made the point that post-war the British had given captured Enigmas to several emerging nations and told them they could use the machines to send and receive secret messages that could not be decoded by an eavesdropper!

Throughout the 1980's, several key players wrote about Bletchley Park — a rural estate north of London —where the British mounted a huge effort to decode and use captured Enigma messages. As more and more of the story unfolded, Britain chose to publicize and glorify these efforts. Now tourists are allowed to tour Bletchley and see how it was done, even to the extent that rebuilt decoding machines are on display.

Personally I did not interest myself in the Enigma story until a movie entitled *Enigma* was released in 2001. 90% fiction, it was occasionally interspersed with pictures and facts about the Enigmas, their usage by the Germans, and the British decoding effort. After watching the film, I looked up a few things about the Enigma and its decoding, but that was that.

In 2014 the movie *Imitation Game* was released whose hero was Alan Turing. Turing was a mathematical genius whose work on computational theory in the early 1930's showed the theoretical power of computing machines. With cinematic liberties, the movie depicts Turing's involvement in the Bletchley Park Enigma efforts from 1938 to 1941. The movie shows Turing's insistence that another machine (the Turing "bombe") be designed and built to crack the messages, and the importance of a crib (i.e. knowing or suspecting a piece of the plain text and its position in the message). It also shows the importance of Gordon Welchman's diagonal board which used the plugboard's own characteristics to nullify its supposed contribution to the complexity of encoded messages. However, another character in the movie is given credit for the diagonal board and Welchman's name is not mentioned.

A book written in 1983 by Andrew Hodges entitled *Alan Turing: The Enigma* was given credit for inspiring *Imitation Game*. I read some of it and did not understand much of the math but decided to challenge myself to write a program to simulate the Enigma machine. By writing and debugging a simulator, I hoped to develop an understanding of Enigma design and usage and have some insight into what was needed to break its secret messages. And yes, it could also be fun to share the program and send and decode "secret" messages

I used a free C/C++ Embarcadero/Borland compiler and linker which ran in the DOS environment provided through selection of the "Command Prompt" option under "Windows System" on my Windows 10 Dell desktop. I wrote the simulator in the C programming language. I put all the source code in one file. I got a preliminary version of the simulator to run successfully in 2015 and then forgot all about it. But during pandemic year 2020 my "stay at home" time increased I finished the simulator and I started to think about writing a "bombe". It didn't take long to see I did not have the data or even the real desire to simulate the Turing "bombe" as it did not really decode the entire message and its setup seemed very complex. So I decided to write a rather simple brute force computer program using only a crib involving the first 6 characters of the encoded message. See the separate section on the Enigma Crib Match Program for further discussion.

ENIGMA COMPONENTS

<u>Figure 1</u>

Figure 1 above gives a view of major Enigma components which are externally visible.

Set aside on the left are two rotors. The lower rotor has an attached alphabet ring while the upper rotor does not.

On the machine itself from bottom to top we see a plugboard, the keyboard, the lamp display, and three rotors placed in the machine with associated windows. Each window shows a letter from the alphabet ring attached to its respective rotor.

Keyboard

The keyboard is a QWERTZ keyboard with a key for each letter of the alphabet. All letters are uppercase. Some early models of the Enigma had no plugboard so the electronic signal emitted when a key was depressed went directly to the corresponding pin on the right side of the rightmost rotor. That is to say the key "A" signal was sent to pin 1, "B" to pin 2, etc. When plugboards were introduced, the key "A" signal went to slot 1 on the plugboard, "B" to slot 2, etc.

Lamp Display

In the Display area, there is a bulb for each letter in the alphabet. After a letter has been processed by the three rotors and the reflector, an electronic signal is emitted from a pin on the right side of the rightmost rotor. Some early models of the Enigma had no plugboard so the electronic signal went directly to keyboard/ lamps. That is to say that a signal emitted from pin 1 on the right side of the rightmost rotor went to the first slot on the keyboard /lamp assembly. A current flow switch (diodes?) detects the electronic flow direction and the current goes to the bulb.

Plugboard

As used by the Germans in WWII, the plugboard settings were set once a day as per the daily entries in the settings book. One side of the plugboard interfaces the keyboard/lamp assembly while the other side interfaces the right side of the rightmost rotor. Plugboard functionality and cable usage is discussed in the Enigma Functional Flow section.

Rotor

The Enigma machine is supplied with 3 rotors and in operation the machine requires that all 3 rotors be utilized. They can be placed in the machine in any order so there are 6 possible rotor sequences. Rotor functionality and wiring is discussed in the Enigma Functional Flow section. Each rotor model I used in my simulator program is uniquely wired. The models copy the wiring of actual rotors used by the Germans in WWII.

Reflector

Reflector functionality and wiring is discussed in the Enigma Functional Flow section.

Figure 2

Enigma Alphabet Rings

The alphabet rings (see Figure 2 above) have three purposes:

a. make it easier for an operator to set the rotors to the correct daily position and message position.

b. offsetting the external appearing letters from the real setting of the rotor. For example the ring value "F" appearing through the window could actually be positioned above the "Q" position on the rotor.

Observation of the initial window settings by an unauthorized individual during the encoding/decoding process would be of little value unless he also knew the alphabet ring settings.

c. The notch on the alphabet ring is used to force the "carry" or step of the neighboring rotor to the left. Note that the notches on the rings are in fixed positions.

For example, assume the alphabet ring on the rightmost rotor has its notch set so that stepping occurs when the window shows "P" and then transitions to "Q". So if the 3 windows show JDP, and the operator depresses a key (any key), the three windows will then show JEQ which means that the middle rotor has advanced one position. Normally this means that for every 26 advancements of the right rotor the middle rotor will advance by 1 position. And the same can be said regarding the middle rotor and the left rotor, viz. that the left rotor advances 1 position for every 26 advancements of the middle rotor.

However there is an exception which is called double stepping. This occurs when a step of the middle rotor then puts its own notch in stepping position for the leftmost rotor. An example would be JDO JDP JEQ KFR KFS where P to Q on the right rotor causes a middle rotor step and E is now positioned to cause the middle rotor notch to step the left rotor. So with the next keystroke both the left rotor and the middle rotor advance. Note that this can cause some sequential settings to be skipped in mid-message. This is not a problem as exactly the same skipping occurs on encoding and decoding.

ENIGMA FUNCTIONAL FLOW

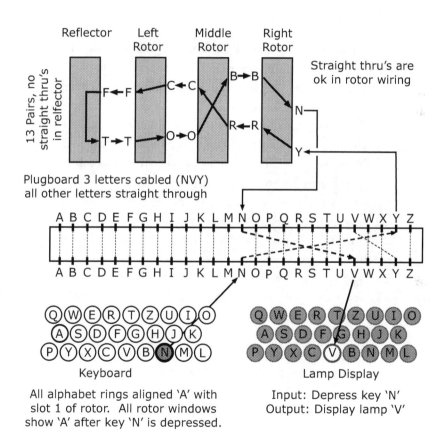

Example showing Enigma functional wiring

Figure 3

Figure 3 above shows the functional flow through the Enigma when a keyboard key is depressed.

The functional flow through the system is as follows:

a. operator depresses a key on the QWERTZ keyboard

b. rotor or rotors are advanced by one logical step

c. electronic flow from keyboard to the keyboard/
    bulb side of plugboard as per depressed key. For
    example, "A" would cause electronic flow to the
    first slot.

d. electronic flow from keyboard/bulb side to
    rotor side depends upon if two different slots on
    respective sides of plugboard have been connected
    by cable (operator normally would do this as per
    daily settings prescribed in the settings book). The
    26 slots on each side of plugboard can be labeled
    with letters of alphabet The cables in use act like
    the internal wiring in a rotor (see below). Note this
    must be a closed set ...Same letters on each side
    are cabled but never to same respective value e.g.,
    "C" to "C" is a no-no, but "C" to "R" could be a
    valid cable connection.

    If, for example, 10 cables are in use, that means
    that 10 letters are "rotor wired" and the remaining
    16 are straight through .. for example an uncabled
    "X" on the keyboard/bulb side will send its signal
    straight across to the uncabled "X" on the rotor
    side.

e. electronic flow from the rotor side of the plugboard
    to the right rotor (right as viewed from the
    operator's position). The flow is straight across in
    that a signal from the "C" slot of plugboard will
    go the third pin position for the right side of the
    right rotor.

The pin positions do not move but as the rotor rotates different slots of the rotor will align and connect to the pin positions. For one setting of the rotor for example the 5th slot on the rotor may be aligned with the 3rd pin position but with the next keystroke and move of the right rotor, the 6th slot on the rotor would be aligned with the 3rd pin position.

f. electronic flow from right side of right rotor to one of 26 positions on the left side of right rotor. Transmission from right to left side of rotor is achieved with 26 internally wired connections. Note that a given connection allows bi-directional flow but never simultaneously). Each connection is unique in that no other connection may use either termination point. Note also that straight-thru connections are allowed.

g. electronic flow from left side of right rotor to right side of middle rotor is straight across, i.e. to same relative pin position on the right side of middle rotor.

h. in essence, do step e and f for middle and right rotors with appropriate word substitution

i. electronic flow out of left side of left rotor to the input side of the reflector is straight across.

j. reflector acts as a fixed rotor with electronic flow from input side to output side with additional constraint that it cannot be to same relative position. This constraint is what allows the left to right electronic flow to use a path which never coincides with the right to left flow.

k. electronic flow from output side of reflector to left side of left rotor is straight across.

l. electronic flow back through the rotors from right to left is just the reverse of the right to left flow.

m. electronic flow from the right side of the right rotor to the rotor side of the plugboard is straight across.

n. flow from the rotor side of the plugboard obeys the pathways as defined by cable or straight across if no cable. However with the current flowing from rotor side to keyboard/bulb side, the pathway opens to light the appropriate bulb.

ENIGMA USAGE

The Enigma Machine is an offline machine where encoding or decoding a message was always a two step sequence. The first step involved using the day settings to either encode or decode the message window settings followed by the second step where day window settings are changed to the message window settings and then the "real" message is either encoded or decoded. Using the identical settings for encoding or decoding worked because at a given setting if one keyed in, for example, an 'A' and the 'N' bulb is illuminated, at the same setting if one keyed in an 'N' the 'A' bulb would be illuminated.

Encoding

A. Encode the Message Window Settings

The first step involved the use of the day settings which were contained in settings books distributed to each Enigma node in the network.

For a given date the settings book specifies

- a. which rotors to use
- b. what order they were to be inserted into the machine
- c. the setting of each rotor's alphabet ring
- d. the beginning window setting of each rotor.

In actual service during the war, the settings book also included the plugboard settings for each day, but my simulator, which uses a plugboard, does not allow those settings to be changed.

To encode a message the operator would configure the machine as per the day settings and would then key in 3 characters defining the message window settings. For example the operator could decide to use WSJ as the message window settings so he would key in WSJ. Hitting the W key for example would cause the plugboard and the scrambler and reflector to encode the W and the encoded value would be shown as a lighted alphabet bulb on the output panel. The operator would then manually write down on paper the letter which had been illuminated. Then he would key in the S, etc., then the J, etc. And then comes the funny twist which helped the codebreakers — the message window settings would be keyed in again. So in this example, the first six characters that would be transmitted via radio would be the encoded values for WSJWSJ.

B. Use the Message Window Settings and Encode the "Real" Message

After encoding the message window settings, per the example settings given above, the operator then manually sets the window settings of the respective rotors to 'W', 'S' and 'J'. Then the operator keys in the "real" message and manually writes down the character shown by the bulb illuminated with each keystroke. After encoding is complete, the handwritten encoded form of the entire message -- i.e. the encoded window message settings and the encoded "real" message -- is then given to the radio operator who transmits the message in Morse code.

Decoding

A. Decode The Message Window Settings

The radio message is received in Morse code and the receiving radio operator manually writes down the alphabetics of the encoded message. The Enigma operator then refers to the settings book to get the day values for the machine rotors, rotor order, alphabet ring settings, and day message settings and he configures the Enigma appropriately. He then keys in the first 6 encoded characters and writes down on paper the letters whose bulbs have been illuminated. As per the example above, the decoded characters should be **WSJWSJ**.

B. Use the Message Window Settings to Decode the "Real" Message

The operator then manually sets the message window settings to the value which he just decoded. Using the above example, he sets the window settings of the respective rotors to 'W', 'S' and 'J'. Then the operator keys in the encoded "real" message and manually writes down the character shown by the bulb illuminated with each keystroke. The written result is the decoded message.

ENIGMA ENCODE AND
DECODE PROGRAM

In discussing the parameters input to my programs, they are expressed in the order of the rotors as seen from the side opposite the operator.

All discussion in this book about left and right rotors are from the view of the operator, left to right. So if the operator is looking at rotors 3 1 2 left to right, my program parameter would say 2 1 3. This is true for rotors, operator ring settings, ring notch settings, and message settings.

The program source code is listed in the Appendix. For personal use only, you can compile it and use the executable object. Note that the program is not guaranteed to be error-free as I only ran 30 or 40 test cases through it.

The program runs in a single directory and only three files are used — the executable object, infile.txt, and outfile.txt. The .txt files are ASCII plain text containing only uppercase alphabetic characters without spaces or any punctuation mark characters. The infile.txt should always be less than 100 characters in size.

infile.txt contains the message to encode/decode and outfile.txt contains the decoded/encoded text.

A plain text infile.txt will contain in the first 6 characters the three letter message settings plus a repeat copy. For example

JSWJSW. This is immediately followed by the plain text of the message to encode.

The program produces the encoded result in the outfile.txt in the same order as contained in infile.txt. Note that the first 6 characters were encoded using the day settings only, while the rest of the message is encoded with message window settings used in place of the day window settings.

With an executable named "enigma"
an example command line for encoding is:

enigma PTE 213 HNR PKL WKO
where enigma is the program name,
PTE tells the program to perform the encode operation, 213 specifies the order of the rotors, HNR specifies the respective settings of the alphabet rings, PKL specifies the respective settings of the notches on the rings, and WKO specifies the respective daily message settings.

My normal way to set up for decoding was to rename and save infile.txt and then rename the encoded outfile.txt as infile.txt.

An example command line for decoding would then be:

> enigma ETP 213 HNR PKL WKO
> where all parameters are the same as in the encode example with the exception of ETP which tells the program to decode an encoded message. The only real difference between the decoding and encoding process involves the message window settings. For encoding, the plain text value in the first 3 characters of infile.txt can be directly used as message window settings. However in the decoding usage, the first 6 characters have to be decoded back to plain text by using the day settings. The first 3 decoded characters are then used for message window settings to decode the rest of the encoded message.

Encoding the below plain text infile.txt using the encoding command in the example above —

JSWJSWFOURSCOREANDSEVENYEARSAGOOURFORE
FATHERSBROUGHTFORTHETC

produces this encoded outfile.txt —

MILNIFLXSWTFMZUCMPWINMYLXFYWOFBRPDJHHCZY
EUOYVOFEFYBSSDIXBPBT

and reversing the process produces the original plain text.

ENIGMA CRIB MATCH PROGRAM

The British efforts to decode encoded Enigma messages sent by the Germans during WWII have in recent years been documented in books and movies. After reading about the efforts at Bletchley Park it was evident that an effort to simulate the Turing Bombe and the Welchman diagonal would be much more difficult than the simulation of the Enigma machine itself. Moreover the output of those efforts was not a completely decoded message but in the best case a crib match with associated settings. This information was then used as setup for enigma machines or simulators at Bletchley Park and subsequent efforts to decode the entire message were often successful.

Of course it should be clear that the primary goal at Bletchley would be to determine the day settings. With the correct day settings, then every encoded message received that day could be decoded. So the effort associated with the early messages of the day were intense.

So let us say it is shortly after midnight and the first encoded messages are arriving. What do the cryptanalysts know? In terms of the settings they know nothing but they know where the message came from (fancy radio direction finding) and they have files on the German radio operators where identifying techniques and past encoded and decoded messages are kept. Let us say that weather messages always come from the same location and usually from the same operator. Perhaps this operator tends to often use the initials of his girlfriend for the message window settings. So the cryptanalyst can use these initials as a crib and set up the

Bombe appropriately to see if on this day the operator reverted to this bad habit. Other messages from known command centers of the German armed forces would often contain the German words for General, Colonel, etc. and they often occurred in the same areas of the messages so these words could also be used for cribs.

Since the output of the Turing Bombe and the Welchman diagonal was a crib match and the effort to write simulators seemed excessive, I thought about writing a computer program which takes a 6 character crib assumed to be two copies of the 3 character encoded message window settings and grinds through all possible Enigma settings until the decoded 6 characters match two concatenated copies of the suspected message window settings. This would be done through repetitive calls of the decode logic contained in my Encode and Decode program discussed and listed in prior pages of this book. That program runs with known values for rotor wiring and alphabet ring notch value, reflector looping, and plugboard cabling.

Before the program was written, simple calculations showed that looking at every possible setting combination would take an enormous amount of execution time. But I wrote the program anyway and thought to learn something by running it to find some pre-planned settings And when I ran the program, I found that literally hundreds of settings can produce the plain text crib in the first 6 characters of the message.

So in practical terms the "everything" program was not ready for prime time.

With a 3 rotor system with three different rotors, there are 6 possible ways they can be configured. And for each configuration there are

676 x 676 x 676 possible settings of alphabet rings and day window settings. On my laptop, testing a settings group which did not produce the crib took about 14 microseconds. To go through a single rotor combination without the occurrence of a match

takes 4300 seconds. So to go through all 6 rotor configurations with minimal matches would take slightly more than 7 hours. If large numbers of matches occur, the recording of the settings will significantly increase processing time.

To reduce execution time without losing matching power, one could possibly ignore alphabet ring settings totally and assume no notch activity in the first few characters of the message. This assumption was often made at Bletchley Park. And this reduced the combinations they worked with to 26 x 26 x 26. For the brute force program discussed above, one could ignore the alphabet ring setting on the left rotor so that the loop count for that rotor would be 26 instead of 676. This would reduce a no match cycle for one rotor configuration from 4300 seconds to 165 seconds and a full minimal match cycle through all 6 rotor combinations would take about 16.5 minutes. And as for presenting the results of a successful match, one could assume the alphabet ring setting of the left rotor was "A" and its day window setting would be the alphabet letter corresponding to the matching rotor slot ... e.g. "D" for a match on the 4th slot.

The crib matching application is tailor-made for parallel processing. A massively parallel approach with 26 computers for each rotor configuration (156 in all) could do a complete scan of the 6 x (26 x 676 x 676) possible settings in less than 10 seconds.

APPENDIX

Below is the source code listing of the Enigma Simulator Program written by Philip Bauer

```
/*                  COPYRIGHT 2021                    */
/*                       by                           */
/*                   Philip Bauer                     */
/*                 All Rights Reserved                */
/* ENIGMA.C */
/* plugboard with 10 cables */
/* stepper notches on alphabet rings */
/* notch logic does double stepping */
/* stepping begins
      with first keystroke after window set ..
*/

#include <stdio.h>
#include <string.h>
#include <ctype.h>
#include <stdlib.h>
#define EQ ==
#define NE !=
#define LT <
#define GT >
#define LE <=
#define GE >=
#define OR ||
#define AND &&
```

```c
/* major parameters for message translation */
int xlatetype; /* 1 = plain to encoded, 2 = encoded to plain */
int rotorsorder[3] ; /* left to right, view opposite operator's */
                    /* left = units, middle = tens, right = hundreds */
char dayringset[4];
char ringnotchset[4]; /* notch is constant on given alphabet ring */
char daywinset[4];
char msgwinset[4];

/* double step work variables */
int ltom,doubleswt;

/* keyslice work area */
int wrkdayring, wrkdaywin, wrkmsgwin;
int wrkslice [26];

/* FILE *iptr; */
/* FILE *optr; */
char *intext = "WSJWSJTHEQUICKBROWNFOXJUMPEDOVER
THELAZYDOGSBACK";
char textout [100];
char outmsg[100];
char inmsg[100];

char xbt [6];
char bt [6];
int inslice;
int outslice;
char inchar;
char outchar;
int msgsize;

char *alphabet = "ABCDEFGHIJKLMNOPQRSTUVWXYZ";
int intalph[26] = {0,1,2,3,4,5,6,7,8,9,10,11,12,13,14,15,16,17,18,
                19,20,21,22,23,24,25};
/* ROTOR WIRING */
char *alphrotc1 = "DMTWSIRLUYQNKFEJCAZBPGXOHV";
```

24

```c
int lrintrotc1 [26]; /* +-offset to connected right pin */
int rlintrotc1 [26]; /* +-offset to connected left pin */

char *alphrotc2 = "HQZGPJTMOBLNCIFDYAWVEUSRKX";
int lrintrotc2 [26] ; /* +-offset to connected right pin */
int rlintrotc2 [26]; /* +-offset to connected left pin */

char *alphrotc3 = "UQNTLSZFMREHDPXKIBVYGJCWOA";
int lrintrotc3 [26]; /* +-offset to connected right pin */
int rlintrotc3 [26]; /* +-offset to connected left pin */

/* REFLECTOR LOOPING */
    char *alphrflec = "IMETCGFRAYSQBZXWLHKDVUPOJN";
    int outpointin [26]; /* keyslice to keyslice */

/* PLUGBOARD WITH 10 CABLES */
char *alphplug = "EBIDCFAHQJKLGNOPVYSTURWXMZ"; /* 10
letters cable-connected */
int lrintplug [26]; /* +- offset to connected right pin */
int rlintplug [26]; /* +- offset to connected left pin */

/* SYSTEM CONFIG */

struct xlatrot { /* for a rotor */
    int rotorid; /* 1 = CI, 2 = CII, 3= CIII */
    int slot; /* 1 = left, 2 = middle, 3 = right */
    int dayringset; /* integer of daily ring set .. e.g.15 for P */
    int daywinset; /* integer of daily window ring letter */
    int msgwinset; /* integer of window ring letter set by operator */

    int keyslice [26]; /* rotor pin nbr for a given key plane */
        /* if day window were set to day ring setting
                        [0] = 0, ...,[25] = 25 */
    /* at start of using window ring letter which e.g. is +2 from day ringset,
    [0] = 2, ..., [24] = 0, [25] =1 */
    /* note ..left is viewed from side opposite of operator ! */
    int lpointr [26]; /* relative+- deltas implied by rotor wiring integers */
    int rpointl [26]; /* inverse of lpointr .. if lpin 0 =+4, then rpin 4 = - 4 */
```

```c
        int stepcounter;
        };

/* temp left rotor work area */
struct xlatrot *wrklft;
struct xlatrot stagelftrot;
/* temp middle rotor work area */
struct xlatrot *wrkmid;
struct xlatrot stagemidrot;
/* temp right rotor work area */
struct xlatrot *wrkrgt;
struct xlatrot stagergtrot;

/* units rotor ... steps with every keystroke */
struct xlatrot rotlft; /* view opposite operator's */

/* tens rotor ..... initial step when notch on units rotor is hit
        and thereafter steps after every 26 character cycle on units rotor */
struct xlatrot rotmid;

/* hundreds rotor ... initial step when notch on tens rotor is hit
        and thereafter steps after every 26 character cycle on tens rotor */
/* double step tens rotor when its kick also causes the 100's rotor to kick */

struct xlatrot rotrgt; /* view opposite operator's */

/* FUNCTION PROTOTYPES FOR MAIN PROCEDURE */
/* prototypes */
        int xlatepluglr (int);
        int xlateplugrl (int);
        int reflectorxlate(int);
        int rotorxlate (int, int, int);
        int xlatechar (int);
        int kickrot (int, int);
        int kicksys(int);
        int xlatechar (int);
        void sliceutil (int);
        void setkeyslice (int, int);
```

```
        void setstepcount (int, int);
        void rotwiredlr (int[], char[]);
        void rotwiredrl (int[], int[]);
        int cnvaton (char);

main (argc, argv)
        int argc;
        char *argv[];
    {
        char *xlatetypeval;
        char *dayrotorvals;
        char *dayringvals;
        char *ringnotchvals;
        char *daywinvals;
        char *msgwinvals;

        char test[4];
        char stest[4];

        FILE *iptr;
        FILE *optr;

        int charcnt;
        char newchar;

        int kk, jj, mm;
        int rotorslot, slicetype;
        char kar1;
        char kar2;
        int intchar;
        int kresult;

/*      get the input parameters and validate them */
        if (argc NE 6)
            {
            printf ("%s \n", "5 input parameters needed, try again");
            return;
            }
```

```
for (kk =1; kk < argc; kk++)
    {
  if (strlen(argv[kk]) NE 3)
    {
    printf (" %s \n", "each param must be 3 chars long, try again");
    return;
    }
}
xlatetypeval = argv[1];
dayrotorvals = argv[2];
dayringvals = argv[3];
ringnotchvals = argv[4];
daywinvals = argv[5];

/* PTE = plain to encoded, ETP = encoded to plain */
strcpy ( test, xlatetypeval); /* must be uppercase alphabetic */
for (kk=0;kk LT 3; kk++)
  {intchar = test[kk];
  if ((intchar LT 65) OR (intchar GT 90))
  { printf (" %s \n", "xlatetype must be uppercase, try again");
      return;
  }
}
xlatetype = 0;
kresult = strcmp ("PTE", xlatetypeval);
if ( kresult EQ 0 ) xlatetype = 1;
kresult = strcmp ("ETP", xlatetypeval);
if (kresult EQ 0 ) xlatetype = 2;
if (xlatetype EQ 0)
    { printf (" %s \n", "bad xlatetypeval, try again");
      return;
    }

/* rotorvalues must be between 1 and 3 */
  strcpy ( test, dayrotorvals);

  for (kk=0;kk LT 3; kk++)
    {
```

```
       intchar = test [kk];
    if ((intchar LT 49) OR (intchar GT 51))
       { printf (" %s \n", "rotor nbr must be 1 thru 3, try again");
         return;
       }
    }
/* int rotorsorder[3] = {1,2,3}; left to right, view opposite operator's */
                  /* left = units, middle = tens, right = hundreds */
    intchar = atoi (test);
    switch (intchar)
    {
    case 123 :
                rotorsorder [0] = 1; /* UNITS */
            rotorsorder [1] = 2; /* TENS */
            rotorsorder [2] = 3; /* HUNDREDS */
            break;
        case 132:
                rotorsorder [0] = 1;
                rotorsorder [1] = 3;
            rotorsorder [2] = 2;
            break;

        case 213:
                rotorsorder [0] = 2;
            rotorsorder [1] = 1;
            rotorsorder [2] = 3;
            break;
        case 231:
                rotorsorder [0] = 2;
            rotorsorder [1] = 3;
            rotorsorder [2] = 1;
            break;
        case 312:
                rotorsorder [0] = 3;
            rotorsorder [1] = 1;
            rotorsorder [2] = 2;
            break;
```

```
        case 321:
                rotorsorder [0] = 3;
            rotorsorder [1] = 2;
            rotorsorder [2] = 1;
            break;
        default:
                printf (" %s \n", "bad rotor nbrs, try again");
                return;
        }


/* char[4] dayringset; */
    strcpy ( test, dayringvals); /* must be uppercase alphabetic */
    for (kk=0;kk LT 3; kk++)
      {intchar = test[kk];
      if ((intchar LT 65) OR (intchar GT 90))
        { printf (" %s \n", "dayringvals must be uppercase, try again");
          return;
        }
      else dayringset[kk] = test[kk];
      }
      dayringset[3] = '\0';


/* char[4] ringnotchvals; */
    strcpy ( test, ringnotchvals); /* must be uppercase alphabetic */
    for (kk=0;kk LT 3; kk++)
      {intchar = test[kk];
      if ((intchar LT 65) OR (intchar GT 90))
        { printf (" %s \n", "ringnotchvals must be uppercase, try again");
          return;
        }
      else ringnotchset[kk] = test[kk];
      }
      ringnotchset[3] = '\0';


/* char[4] daywinset; */
    strcpy ( test, daywinvals); /* must be uppercase alphabetic */
    for (kk=0;kk LT 3; kk++)
      {intchar = (test[kk]);
```

```c
        if ((intchar LT 65) OR (intchar GT 90))
           { printf (" %s \n", "daywinvals must be uppercase, try again");
              return;
           }
           else daywinset[kk] = test[kk];
        }
        daywinset[3] = '\0';

printf(" xlatetype = %i \n", xlatetype);
printf (" rotor sub 0 is units, sub 1 is tens, sub 2 is hundreds \n");
for (kk = 0; kk < 3; kk++)
     printf (" rotors sub %i = %i \n", kk, rotorsorder[kk]);
printf (" day ring sets = %s \n", dayringset);
printf (" ring notch values = %s \n", ringnotchset);
printf (" day window sets = %s \n", daywinset);

/* get the file to translate */
     iptr = fopen ("infile.txt", "r");
     charcnt = 0;

     while (!feof(iptr))
     {
       newchar = fgetc (iptr);
       intchar = newchar;
       /* accept only uppercase alphabetic characters */
       if (intchar GE 65 ) /* A */
          {
          if (intchar LE 90 ) /* Z */
             {
                intext [charcnt] = newchar;
                intext [charcnt+1] = '\0';
                charcnt++;
             }
          }
     }
fclose (iptr);

printf ("input file size = %i \n", charcnt);
```

```c
/* if input msg is plaintext, first 3 chars are msg window settings */
/* next 3 are a repeat of first 3, but we are not going to verify this */
    if (xlatetype EQ 1)
        {
            for (jj = 0; jj < 3; jj++) msgwinset[jj] = intext[jj];
            msgwinset [3] = '\0';
            printf (" msg window sets = %s \n", msgwinset);
        }
/* set reflector array in terms of offsets from A */
/* A is 41 hex */
    for (jj = 0; jj < 26; jj++)
        { kk = alphrflec [jj]; outpointin[jj] = kk - 65;}


/* plugboard is essentially a fixed rotor */
/* set up the plugboard left-to-right and right-to-left arrays */
    rotwiredlr (lrintplug, alphplug);
    rotwiredrl (rlintplug, lrintplug);


/* set up the l-to-r and r-to-l arrays for rotors C1, C2, and C3 */
    rotwiredlr (lrintrotc1, alphrotc1);
    rotwiredrl (rlintrotc1, lrintrotc1);

    rotwiredlr (lrintrotc2, alphrotc2);
    rotwiredrl (rlintrotc2, lrintrotc2);

    rotwiredlr (lrintrotc3, alphrotc3);
    rotwiredrl (rlintrotc3, lrintrotc3);
/* fill the rotor structures with constants and calculated arrays */

/* left rotor */
    rotlft.rotorid = 1;
    rotlft.slot = 1;
    rotlft.stepcounter = 0;

/* middle rotor */
    rotmid.rotorid = 2;
    rotmid.slot = 2;
    rotmid.stepcounter = 0;
```

```
/* right rotor */
    rotrgt.rotorid = 3;
    rotrgt.slot = 3;
    rotrgt.stepcounter = 0;

    /* set pointers for moves to staging areas */
    for (kk = 0; kk < 3; kk++)
     { switch (kk) {
     case 0: /* left rotor */
          jj = rotorsorder [0];
          if (jj EQ 1) wrklft = &rotlft;
          if (jj EQ 2) wrklft = &rotmid;
          if (jj EQ 3) wrklft = &rotrgt;
          break;
      case 1: /* middle rotor */
          jj = rotorsorder [1];
          if (jj EQ 1) wrkmid = &rotlft;
          if (jj EQ 2) wrkmid = &rotmid;
          if (jj EQ 3) wrkmid = &rotrgt;
          break;

     case 2: /* right rotor */
          jj = rotorsorder [2];
          if (jj EQ 1) wrkrgt = &rotlft;
          if (jj EQ 2) wrkrgt = &rotmid;
          if (jj EQ 3) wrkrgt = &rotrgt;
          break;

      }
}
/* move the rotor structures to correct places in staging area */
stagelftrot.rotorid = (*wrklft).rotorid;
stagelftrot.slot = 1 ;
stagelftrot.stepcounter = 0;
stagelftrot.dayringset = cnvaton (dayringset[0]) ;
stagelftrot.daywinset = cnvaton(daywinset[0]) ;
stagelftrot.msgwinset = cnvaton (msgwinset[0]);
```

```
rotlft.rotorid = stagelftrot.rotorid;
rotlft.slot = stagelftrot.slot;
rotlft.dayringset = stagelftrot.dayringset ;
rotlft.daywinset = stagelftrot.daywinset ;
rotlft.msgwinset = stagelftrot.msgwinset ;
rotlft.stepcounter = stagelftrot.stepcounter;

stagemidrot.rotorid = (*wrkmid).rotorid;
stagemidrot.slot = 2;
stagemidrot.stepcounter = 0 ;
stagemidrot.dayringset = cnvaton (dayringset[1]) ;
stagemidrot.daywinset = cnvaton(daywinset[1]) ;
stagemidrot.msgwinset = cnvaton (msgwinset[1]);

rotmid.rotorid = stagemidrot.rotorid;
rotmid.slot = stagemidrot.slot;
rotmid.dayringset = stagemidrot.dayringset ;
rotmid.daywinset = stagemidrot.daywinset ;
rotmid.msgwinset = stagemidrot.msgwinset ;
rotmid.stepcounter = stagemidrot.stepcounter;

stagergtrot.rotorid = (*wrkrgt).rotorid;
stagergtrot.slot = 3;
stagergtrot.stepcounter = 0;
stagergtrot.dayringset = cnvaton (dayringset[2]) ;
stagergtrot.daywinset = cnvaton(daywinset[2]) ;
stagergtrot.msgwinset = cnvaton (msgwinset[2]);

rotrgt.rotorid = stagergtrot.rotorid;
rotrgt.slot = stagergtrot.slot;
rotrgt.dayringset = stagergtrot.dayringset ;
rotrgt.daywinset = stagergtrot.daywinset ;
rotrgt.msgwinset = stagergtrot.msgwinset ;
rotrgt.stepcounter = stagergtrot.stepcounter;
```

```
switch (rotlft.rotorid) {
    case 1:
        for (jj= 0; jj <26; jj++)
            { rotlft.lpointr[jj] = lrintrotc1[jj];
                rotlft.rpointl[jj] = rlintrotc1[jj];
            }
        break;
    case 2:
        for (jj= 0; jj <26; jj++)
            { rotlft.lpointr[jj] = lrintrotc2[jj];
                rotlft.rpointl[jj] = rlintrotc2[jj];
            }
        break;
    case 3:
        for (jj= 0; jj <26; jj++)
            { rotlft.lpointr[jj] = lrintrotc3[jj];
                rotlft.rpointl[jj] = rlintrotc3[jj];
            }
        break;
    }

switch (rotmid.rotorid) {
    case 1:
        for (jj= 0; jj <26; jj++)
            { rotmid.lpointr[jj] = lrintrotc1[jj];
                rotmid.rpointl[jj] = rlintrotc1[jj];
            }
        break;
    case 2:
        for (jj= 0; jj <26; jj++)
            { rotmid.lpointr[jj] = lrintrotc2[jj];
                rotmid.rpointl[jj] = rlintrotc2[jj];
            }
        break;
```

```
    case 3:
        for (jj= 0; jj <26; jj++)
            { rotmid.lpointr[jj] = lrintrotc3[jj];
                rotmid.rpointl[jj] = rlintrotc3[jj];
            }
        break;
    }

switch (rotrgt.rotorid) {
    case 1:
        for (jj= 0; jj <26; jj++)
            { rotrgt.lpointr[jj] = lrintrotc1[jj];
                rotrgt.rpointl[jj] = rlintrotc1[jj];
            }
        break;
    case 2:
        for (jj= 0; jj <26; jj++)
            { rotrgt.lpointr[jj] = lrintrotc2[jj];
                rotrgt.rpointl[jj] = rlintrotc2[jj];
            }
        break;
    case 3:
        for (jj= 0; jj <26; jj++)
            { rotrgt.lpointr[jj] = lrintrotc3[jj];
                rotrgt.rpointl[jj] = rlintrotc3[jj];
            }
        break;
    }

slicetype = 1; /* i.e., using day window settings */
for (rotorslot =1; rotorslot < 4; rotorslot++)
    {
        setkeyslice (rotorslot, slicetype);
    /* now set stepcounters relative to day window settings */
        setstepcount (rotorslot, slicetype);

    }
```

```c
mm = 0; /* count of kicksys calls */
ltom = 0; doubleswt = 0; /* set double step variables to off */

/*      now ready to translate the message */
   for (kk = 0; kk < 6; kk++)
     {

        mm =kicksys(mm);

        inslice = intext [kk] - 65; /* offset from A which is hex 41 */
        bt[kk] = intext [kk];

        outslice = xlatepluglr (inslice);
        inslice = outslice;

        outslice = xlatechar (inslice);

        inslice = outslice;
        outslice = xlateplugrl (inslice);

        outchar = outslice + 65;
        xbt[kk] = outchar;
        textout[kk] = outchar;
     }

/* if input= encoded, then first 3 xlated chars are windowsets */
if (xlatetype EQ 2)
     {
        for (jj = 0; jj < 3; jj++) msgwinset[jj] = xbt[jj];
        msgwinset [3] = '\0';
        printf (" msg window sets = %s \n", msgwinset);

        rotlft.msgwinset = cnvaton (xbt[0]);
        rotmid.msgwinset = cnvaton (xbt[1]);
        rotrgt.msgwinset = cnvaton (xbt[2]);
     }
```

```c
/*      set keyslice for message settings */
    slicetype = 2; /* 1 = day only, 2 = message */
/* rotorslot = 1 = left, 2 = middle, 3 = right */
    for (rotorslot = 1; rotorslot < 4; rotorslot++)
        {
            setkeyslice ( rotorslot, slicetype );
            /* now set stepcounters relative to msg window settings */
            setstepcount (rotorslot, slicetype);
        }

msgsize = strlen (intext);
printf ("msgsize = %i \n", msgsize);

ltom = 0; doubleswt = 0; /* reset double step variables to off */

    for (kk = 6; kk < msgsize; kk++)
        {
            mm = kicksys(mm);
            inslice = intext [kk] - 65; /* offset from A which is hex 41 */

            outslice = xlatepluglr (inslice);
            inslice = outslice;

            outslice = xlatechar (inslice);

            inslice = outslice;
            outslice = xlateplugrl (inslice);

            outchar = outslice + 65;
            textout[kk] = outchar;
            textout[kk+1] = '\0'; /* end of string */
        }

printf ("translated msg is as follows: \n%s \n" , textout);
printf (" translated msg size = %i \n", mm);
```

```c
/* now write the outfile */
optr = fopen ("outfile.txt", "w");

for (charcnt = 0; charcnt <mm ; charcnt++)
    {
        newchar = textout[charcnt];
        fputc (newchar, optr);
    }
  newchar = '\0';
  fputc (newchar, optr);
fclose (optr);

return;
}

/* FUNCTIONS */

  int cnvaton ( kar)
      char kar;
  {
      int aint, karint, delta;
      aint = 'A';
          karint = kar;
      delta = karint - aint;
      if (delta LT 0) return -1;
      if (delta GT 25) return -1;
      return delta;
  }

      void rotwiredlr (result26, wd26)
      int result26[];
      char wd26[];
  {
      int jj, kk;
      for (jj = 0; jj < 26; jj++)
        {
            kk = cnvaton (wd26[jj]);
```

```
                result26[jj] =kk - jj;
            }
    }


    void rotwiredrl (rlresult26, lrinput26)
    int rlresult26[];
    int lrinput26[];
    {
    int rightsub, kk;
    for (kk = 0; kk < 26; kk++)
        {
            rightsub = lrinput26[kk];
            rightsub = kk +rightsub;
            rlresult26 [rightsub] = 0;
            if (lrinput26[kk] NE 0)
                rlresult26 [rightsub] = -lrinput26[kk];
        }
    }


    void setkeyslice (rotslot, ktype )
            int rotslot; /* 1 = left, 2 = middle, 3 = right */
            int ktype; /* 1 = day settings, 2 = day ring, oper window */
    {
            int k;
            switch (rotslot)
            {
            case 1:
                wrkdayring = rotlft.dayringset;
                wrkdaywin = rotlft.daywinset;
                wrkmsgwin = rotlft.msgwinset;
                sliceutil (ktype);
                for (k = 0; k < 26; k++) rotlft.keyslice[k] = wrkslice[k];
                break;
            case 2:
                wrkdayring = rotmid.dayringset;
                wrkdaywin = rotmid.daywinset;
                wrkmsgwin = rotmid.msgwinset;
                sliceutil (ktype);
```

```c
            for (k = 0; k < 26; k++) rotmid.keyslice[k] = wrkslice[k];
          break;
       case 3:
            wrkdayring = rotrgt.dayringset;
            wrkdaywin = rotrgt.daywinset;
            wrkmsgwin = rotrgt.msgwinset;
            sliceutil (ktype);
            for (k = 0; k < 26; k++) rotrgt.keyslice[k] = wrkslice[k];
          break;
       }
   }

void sliceutil (type)
       int type; /* 1 = day, 2 = msg */
   {
     int workval;
     int workwin;
     int k;

     if (type EQ 1) /* day */
        workwin = wrkdaywin;
     else /* msg */
        workwin = wrkmsgwin;

     workval = workwin - wrkdayring;
     if (workval >= 0) wrkslice[0] = workval;
     else wrkslice[0] = workval + 26;
     for (k = 0; k < 25; k++)
        {
        wrkslice [k + 1] = wrkslice [k] + 1;
        if (wrkslice [k + 1] > 25) wrkslice [k + 1] = 0;
        }
   }
```

```
void setstepcount (rotslot, ktype )
     int rotslot; /* 1 = left, 2 = middle, 3 = right */
     int ktype; /* 1 = day settings, 2 = day ring, oper window */
  {
  int wrk;
  switch (rotslot)
     {
       case 1:
            /* assume ktype = 1 viz., day settings */
              wrk = cnvaton (ringnotchset[0]) - rotlft.daywinset;
              if (ktype EQ 2) /* message settings */
              {
                 wrk = cnvaton(ringnotchset[0]) - rotlft.msgwinset;
              }
            if (wrk LT 0) wrk = -wrk;

              rotlft.stepcounter = wrk;
              break;

       case 2:
            /* assume ktype = 1 viz., day settings */
              wrk = cnvaton(ringnotchset[1]) - rotmid.daywinset;
              if (ktype EQ 2) /* message settings */
              {
                 wrk = cnvaton(ringnotchset[1]) - rotmid.msgwinset;
              }
            if (wrk LT 0) wrk = -wrk;

              rotmid.stepcounter = wrk;
              break;

       case 3:
            /* assume ktype = 1 viz., day settings */
              wrk = cnvaton(ringnotchset[2]) - rotrgt.daywinset;
              if (ktype EQ 2) /* message settings */
              {
                 wrk = cnvaton(ringnotchset[2]) - rotrgt.msgwinset;
              }
```

```
                if (wrk LT 0) wrk = -wrk;

                rotrgt.stepcounter = wrk;
                break;
                }
            return;
        }


int kicksys(jmn)
    int jmn;
    {
    int runtot;
    int trigger; /* 0 = no kick, 1 = kick */
    int slotrot; /* rotor slot, 1 == left, ..., 3 = right*/

    trigger = 1; /* left rotor always kicks */
    for (slotrot = 1; slotrot < 4; slotrot++)
                    trigger = kickrot (trigger, slotrot);
    runtot = jmn + 1;
    return runtot;
    }


int kickrot (trig, slotno)
    int trig; /* 0 = no kick, 1 = kick */
    int slotno; /* 1 = left, ..., 3 = right */
    {
    int n;
    int wtrig;
    wtrig = 0;
    if (trig EQ 0) return wtrig;

    switch (slotno)
    {
      case 1:
        for (n=0; n < 26; n++)
          {
            rotlft.keyslice[n]++;
```

```
        if (rotlft.keyslice[n] > 25) rotlft.keyslice[n] = 0;
    }
    rotlft.stepcounter++;
    if (rotlft.stepcounter > 25)
    {
        rotlft.stepcounter = 0;
        wtrig = 1;
        ltom = 1; /* indicates middle kick on for dbl step logic */
    }
    else /* is double stepping needed */
    {
        if (doubleswt EQ 1)
        {
        ltom = 0;
            doubleswt = 0;
        wtrig =1; /* double step will happen */
        }
    }
    break;

case 2:
    for (n=0; n < 26; n++)
    {
        rotmid.keyslice[n]++;
        if (rotmid.keyslice[n] > 25) rotmid.keyslice[n] = 0;
    }
    rotmid.stepcounter++;
    if (rotmid.stepcounter > 25)
    {
        rotmid.stepcounter = 0;
        wtrig = 1;
        if (ltom EQ 1) doubleswt = 1; /* turn on double kick if needed */
    }
    else ltom = 0; /* no doublestep, be sure all indicators cleared */
break;
```

```
        case 3:
          for (n=0; n < 26; n++)
            {
              rotrgt.keyslice[n]++;
              if (rotrgt.keyslice[n] > 25) rotrgt.keyslice[n] = 0;
            }
          rotrgt.stepcounter++;
          if (rotrgt.stepcounter > 25)
            {
              rotrgt.stepcounter = 0;
              wtrig = 1;
            }
          break;

        }
      return wtrig;
}

int xlatechar (nslice )
      int nslice;

{

      int inslice;
      int outslice;
      int rotslt; /* rotor slot 1=left, 2=mid,3=right */
      int direction;
      int k;

      inslice = nslice;
      direction = 1; /* 1 = left to right, 2 = right to left */

      for (rotslt = 1; rotslt < 4; rotslt++)
        {
          outslice = rotorxlate (direction, rotslt,inslice);
            inslice = outslice;
        }
```

```
/* inslice is set correctly */
    outslice = reflectorxlate (inslice);
    inslice = outslice;
    direction = 2;
    for (rotslt = 3; rotslt > 0; rotslt--)
      {
        outslice = rotorxlate (direction, rotslt, inslice);
        inslice = outslice;
      }
    return outslice;
}

int rotorxlate (dir,rslot,inslyce)
    int dir; /* direction 1 = left to right, 2 = right to left */
    int rslot; /* rotor slot 1 = left, ..., 3 = right */
    int inslyce; /* keyslice entry to side implied by direction */
    /* this has lazy inefficient coding .. speed it up later */
    /* this function returns keyslice of exit from other side of rotor */
  {
    int nn;
    int leftright[26];
    int rightleft[26];
    int keyslyce[26];
    int pinnbr; /* zero origin */
    int workval;

    switch (rslot) {
        case 1 : /* left rotor */
          for (nn = 0; nn < 26; nn++)
            {
                leftright[nn] = rotlft.lpointr[nn];
                rightleft[nn] = rotlft.rpointl[nn];
                keyslyce[nn] = rotlft.keyslice[nn];
            }
          break;
```

```
          case 2: /* middle rotor */
             for (nn = 0; nn < 26; nn++)
               {
                  leftright[nn] = rotmid.lpointr[nn];
                  rightleft[nn] = rotmid.rpointl[nn];
                  keyslyce[nn] = rotmid.keyslice[nn];
               }
             break;

          case 3: /* right rotor */
             for (nn = 0; nn < 26; nn++)
               {
                  leftright[nn] = rotrgt.lpointr[nn];
                  rightleft[nn] = rotrgt.rpointl[nn];
                  keyslyce[nn] = rotrgt.keyslice[nn];
               }
             break;
        }
pinnbr = keyslyce [inslyce];
if (dir EQ 1 ) /* left to right */
      workval = inslyce + leftright[pinnbr];
else workval = inslyce + rightleft[pinnbr]; /* right to left */

if (workval > 25 ) workval = workval -26;
if (workval < 0 ) workval = workval +26 ;
/* workval is now the exiting keyslice */
return workval;

}

int reflectorxlate(exitslice)
    int exitslice; /* keyslice out the right side of rightmost rotor */
{
    int entryslice; /* keyslice entering right side of rightmost rotor */

    entryslice = outpointin [exitslice];
    return entryslice;
}
```

```c
int xlatepluglr (entryslice)
    int entryslice; /* keyslice entering left side of plugboard */
    {
    int exitslice; /* keyslice exiting right side of plugboard */

    exitslice = entryslice + lrintplug [entryslice];
    return exitslice;
    }

int xlateplugrl (entryslice)
    int entryslice; /* keyslice entering right side of plugboard */
    {
    int exitslice; /* keyslice exiting left side of plugboard */

    exitslice = entryslice + rlintplug [entryslice];
    return exitslice;
    }
    □
```

Printed in the United States
by Baker & Taylor Publisher Services